WHAT DOES IT LOOK LIKE?

IT HAS FINS!

BY KATHLEEN CONNORS

Gareth Stevens PUBLISHING

first concepts

Some animals have fins.
Most fish have fins.

Angelfish have fins.

Clown fish have fins.

Tuna have fins.

Sharks are fish.
They have fins.

Seahorses are fish too.
They have fins.

Other water animals have fins.

Whales have fins.

Dolphins have fins.

Some things have fins.
Airplanes have fins.

Surfboards have fins.

Which animal has fins?

Please visit our website, www.garethstevens.com. For a free color catalog of all our high-quality books, call toll free 1-800-542-2595 or fax 1-877-542-2596.

Cataloging-in-Publication Data

Names: Connors, Kathleen.
Title: It has fins! / Kathleen Connors.
Description: Buffalo, NY : Gareth Stevens Publishing, 2026. | Series: What does it look like?
Identifiers: ISBN 9781482470840 (pbk.) | ISBN 9781482470857 (library bound) | ISBN 9781482470864 (ebook)
Subjects: LCSH: Fins (Anatomy)–Juvenile literature. | Flippers (Anatomy)–Juvenile literature. Fishes–Juvenile literature. | Aquatic animals–Juvenile literature.
Classification: LCC QL950.75 C747 2026 | DDC 591.47'9–dc23

Published in 2026 by
Gareth Stevens Publishing
2544 Clinton Street
Buffalo, NY 14224

Designer: Andrea Davison-Bartolotta
Editor: Kristen Nelson

Photo credits: Cover, p. 1, series art (background) GraphicsRF.com/Shutterstock.com; cover, p. 1 (fish) Lotus_studio/shutterstock.com; p. 3 (top) frantisekhojdysz/Shutterstock.com; p. 3 (bottom) Victor1153/Shutterstock.com; p. 5 Joan Carles Juarez/Shutterstock.com; p. 7 Kurit afshen/Shutterstock.com; p. 9 Nicolas-SB/Shutterstock.com; p. 11 kaschibo/Shutterstock.com; p. 13 GOLFX/Shutterstock.com; p. 15 Jonas Gruhlke/Shutterstock.com; p. 17 Tara Lambourne/Shutterstock.com; p. 19 Synthetic Messiah/Shutterstock.com; p. 21 moonmovie/Shutterstock.com; p. 23 (cat) Le Dang Nhut Minh/Shutterstock.com; p. 23 (goldfish) Vangert/Shutterstock.com; p. 23 (porcupine) Eric Isselee/Shutterstock.com; p. 23 (turtle) Sutthiphong Chandaeng/Shutterstock.com.

Printed in China

CPSIA compliance information: Batch #QSGS26: For further information contact Gareth Stevens at 1-800-542-2595.